ACTION EVENTS

TRUCK AND TRACTOR PULLERS

ACTION EVENTS

TRUCK AND TRACTOR PULLERS

BY JEFF SAVAGE

CRESTWOOD HOUSE
Parsippany, New Jersey

Photo Credits
(all photos) United States Hot Rod Association.

Cover and book design by Liz Kril

Published by Crestwood House,
A Division of Simon & Schuster
299 Jefferson Road, Parsippany, NJ 07054

First Edition
Printed in the United States of America
10 9 8 7 6 5 4 3 2 1

Library of Congress Cataloging-in-Publication Data
Savage, Jeff, 1961-
Truck and tractor pullers/by Jeff Savage.—1st ed.
p. cm.—(Action events)
Includes index.

Summary: Traces the history of truck and tractor pulling from its beginning in the Midwest in the late 1800s and includes coverage of famous drivers and exciting events.

ISBN 0-89686-886-9 (lib. bdg.)—ISBN 0-382-39296-5 (pbk.)

1. Tractor driving—Competitions—Juvenile literature. [1. Truck driving—Competitions. 2. Tractor driving—Competitions.] I. Title. II. Series:
Savage, Jeff, 1961- Action events.
TL233.3.S278 1997
796.7—dc20 95-35532

AND THE
TEAM

Contents

Chapter 1

Caught by the Bug

Jim Brockman was minding his equipment rental shop one day when the telephone rang. It was a local promoter. Jim was told that a tractor-pulling event would be taking place that weekend in Jim's hometown of New Haven, Indiana. The promoter needed a tractor and grader to smooth the dirt track. He wanted to know if he could borrow the equipment from Jim. In return, Jim would get free advertising for his shop.

"I'll loan you the equipment on one condition," Jim said to the promoter. "I'll do it if you let me drive the tractor at the event. I can smooth dirt as well as anybody."

The promoter agreed to the deal. Jim smiled and hung up the phone. He was happy for two reasons—he would get in free to the event and he would be close to the action. At the time, Jim didn't know much about tractor pulling. But that was about to change.

The weekend arrived and Jim showed up at the track with his equipment. Fans were already filtering into the stadium. The show would be starting soon. The promoter explained to Jim how to keep the dirt even and solid. Jim learned

when to drive his tractor onto the track and where the most important areas to tend were. It was more complicated than he thought. By this time, thousands of people had filled the stadium stands. It was a warm summer evening, perfect weather for a tractor pull.

Jim was excited as he waited for the event to start. Two tractors lined up side by side, each with a giant red sled hitched to the back. The stadium announcer introduced the competitors, and the crowd cheered. The drivers revved their engines, the green flag was waved, and the tractors roared forward. Jim couldn't believe it. The noise was thunderous!

THE SOUND OF A TRUCK AND TRACTOR PULL IS THUNDEROUS.

The engines roared as the tractors surged ahead. Their huge rear tires spit dirt into the stands. Smoke spewed into the air like twin geysers. The crowd cheered with delight. As the tractors rumbled down the track, Jim could feel the engine vibrations rivet through his body. He was so stunned that when the race ended, he almost forgot to drive his tractor out to smooth the track.

The evening continued this way. Dust blew in Jim's face. Dirt soiled his clothes. He was right in the middle of the action. It was all thrilling to him.

"That's when I got caught by the bug," Jim remembered. For on that evening back in 1973, the course of Jim Brockman's life changed forever. He decided then and there to become a professional tractor puller.

Jim's favorite competitor that night was Lynn Hoffman. After the event, Jim asked to join Lynn's crew and was welcomed. As the months passed, the two men developed a close friendship.

Jim learned all he could from Lynn and soon felt ready to compete on his own. He built a powerful tractor engine with Lynn's help and began racing it.

But he wasn't as good as some of the veteran drivers. Also, he found it to be quite expensive to haul his racing tractor from event to event. Lynn had the same problem. So the two friends decided to become racing partners and travel to the events together.

In 1980 the team became a family affair when Jim married Lynn's sister, Cheryl. Jim and Lynn became known as the Inlaws and Outlaws Pulling Team. And Jim began winning.

In 1982 he won the SRO (Standing Room Only) points championship in two different classes—the 1,500 **minirods** and the 1,750 minirods. He won the 1,500 minirods title again the following year. In 1991 he won another championship in a class known as the 7,200 **modified** tractor division.

Lynn stopped racing a few years ago, leaving Jim as the main force behind Inlaws and Outlaws. He competes now with two vehicles—a modified tractor and a **two-wheel-drive** truck. Most pulling events around the country feature both types of vehicles, often at the same time. Jim sometimes races his tractor down one track and then sprints across the field to race his truck down another track. "It's kind of difficult watching two tracks at once," he says. "It gets your adrenaline flowing."

Jim doesn't mind running back and forth. He's out there having fun—and winning, too. After 20 years of racing, Jim is one of the most talented and respected truck and tractor pullers around. And it's all because he spoke up on the phone one day and asked a promoter if he could smooth the dirt track at a local event.

TRUCKS PULL THE SLED WITH SUCH FORCE THAT THEY ARE OFTEN LIFTED INTO THE AIR.

CHAPTER 2

IN THE BEGINNING

Truck and tractor pulling is one of the fastest growing motorsports today. The United States Hot Rod Association (USHRA) sponsors a series of races throughout the country that thousands of people attend each week. Other local events draw large crowds as well. Huge stadiums like the Astrodome in Houston and Busch Stadium in St. Louis have been jam-packed with tractor- and truck-pulling fans. More than 70,000 people have filled the Silverdome in Pontiac, Michigan, to see a pulling event. Thousands of people tune in weekly to watch the sport on the television sports network ESPN. There are fans stretching from the East Coast to the West Coast and everywhere in between. But truck and tractor pulling is especially popular in the region known as the Heartland of America. This is the farmland country of the Midwest. And this is where tractor pulling grew from humble beginnings.

Back in the late 1800s, farm animals were used to pull plows and wagons. Farmers sometimes held contests to see whose horse or oxen could pull the most weight. In the early 1900s, machines began to replace the animals. Tractors were used to plow the fields. The stronger the tractor, the faster and deeper the plow could go. Farmers proudly displayed their new machinery at local county fairs and

THE UNITED STATES HOT ROD ASSOCIATION (USHRA) PROVIDES ALL THE EQUIPMENT FOR COMPETITION — EXCEPT THE PULLERS.

VINTAGE TRUCKS WITH MODIFIED ENGINES ARE AMONG TODAY'S PULLERS.

engaged one another in various pulling competitions. These events were entertaining to farmers and onlookers alike.

By the 1930s, tractor-pull tournaments known as **tug pulls** were being held throughout the Midwest. In a tug pull a sheet of steel was loaded with rocks, concrete slabs, or other heavy objects and then hitched to the back of a tractor. If the tractor could pull the weight ten feet, it would qualify for the next round. If not, it was out of the contest. For the next round, more rocks and concrete slabs were loaded onto the steel sheet. The competition continued for as many rounds as it took until only one tractor remained that could pull the weight ten feet. That tractor was declared the champion.

In the 1960s the steel sheet was replaced by a **step-on sled**. Men would line up along the track ten feet apart. Then as the tractor came down the track, the men would step, one by one, onto the sled, adding their weight to it. More men stepped onto the sled as the tractor continued down the track until finally it could no longer haul the weight and came to a stop. The tractor that had traveled the farthest was declared the winner.

Drivers also began to compete in pulling contests with their trucks. They used step-on sleds, just as with the tractors. Drivers continued to improve their engines by adding more horsepower. By the 1970s, tractors and trucks were so powerful that the step-on sled became hazardous. A new sled was created that is still used today.

Chapter 3

Today's Pullers

The sled pulled by trucks and tractors today is an example of progress. It is a modern device known as a **weight transfer machine**. This sled looks a lot like a truck trailer. It has a long flat bed with tires in the back and weighs about one ton. In the front of the sled is a piece of steel known as a **skid plate**. This plate hovers just above the ground at the start of the race. But as the sled is pulled down the track, the skid plate slowly drops until it digs into the ground. This occurs as a result of the shifting of 15,000 pounds of movable weights on the sled!

The weights are over the rear tires at the start of the pull, and there is no pressure on the skid plate up front. As the sled is pulled forward, the weights are automatically moved forward as well—up the long flat bed toward the front. With more and more weight coming toward it, the steel plate sinks lower and lower until finally it skids along the ground. As the truck or tractor works to overcome this resistance, the plate sinks deeper into the ground. It takes a powerful puller to reach the finish line hauling all that weight *and* a steel plate digging into the ground. Making it to the finish line is called a **full pull**. If two or more drivers accomplish a full pull in the same contest, a **pull off** takes place. This is an event in which the pullers who have gone the distance the first time battle it out again—this time with more weight added to the sled.

A sled operator makes sure the machine is working correctly. He has a **"kill" switch** that allows him to shut off the puller's engine in an emergency. And where is the sled operator posi-

THE SLED PULLED TODAY IS A MASSIVE DEVICE KNOWN AS THE WEIGHT TRANSFER MACHINE.

tioned to get the best view of the sled and the puller? It's not at the starting line or the finish line. The operator actually stands on the sled during the pull!

A sled usually is pulled farther by a tractor than a truck. This is because a truck is permitted to have only one engine. A tractor can use as many engines as the driver wants. And the engines can be any size. The only limit for a tractor is a total weight not over 7,200 pounds. That's room for a lot of engines!

UNLIMITED DRAGSTERS CAN BE POWERED BY AS MANY ENGINES AS WILL FIT ON THE VEHICLE.

There are two main types of trucks—two-wheel-drive and **four-wheel-drive**. A four-wheel-drive truck provides power from the engine to all four wheels. A two-wheel-drive truck is powered only by its rear wheels. You can tell the difference between the trucks by the size of the back tires. A two-wheel-drive truck's back tires are gigantic!

The average pulling vehicle, whether it's a tractor or a truck, costs nearly $200,000. And each trip down a 300-foot track uses more than four gallons of fuel. It's a good thing for the drivers that there's plenty of prize money. In 1994 alone, pullers competed for more than $1 million in cash and prizes.

Chapter 4

Building a Track

"The first time I saw a tractor pull—oh, man, I was hooked," says tractor pull driver John Powell. "Just pure raw **horsepower**."

John lives near Raleigh, North Carolina, and drives the powerful tractor Rolling Thunder. This tractor has four engines at 2,000 horsepower each. That's about the same as two train engines! "The tires are turning fast enough to go about 100 miles per hour if you cut the tractor loose from the sled," John says. "And indoors, most of the time you're running toward a concrete wall. You better be hooked up good."

A secure **hitch** is vital in truck and tractor pulling. But just as important, as John and the other drivers know, is the surface of the track. A pulling event cannot take place just anywhere. A solid dirt pack is needed for the tires to get traction and the sled to dig into the ground. This is where an expert like Randy Spraggins is needed. Randy is the main track builder for the USHRA. It's his job to provide a good quality track for the competitors.

Randy first has to find the dirt. He either buys it or borrows it. In many cities he goes on a dirt hunt to find the right type of dirt. It should be dry and without rocks, pebbles, or sand. A rock shooting out from a spinning tire and flying into the stands is extremely dangerous. The best dirt is black dirt. The worst is red dirt. Once Randy finds his dirt, he must rent or borrow the heavy equipment needed to move the dirt to the stadium or arena.

Sometimes Randy's crew has less than one day to build the track. Bulldozers, loaders, rollers, grading tractors, and backhoes are the machines used to properly spread the dirt. Watching the construction of the track can sometimes be as thrilling as the pulling event itself.

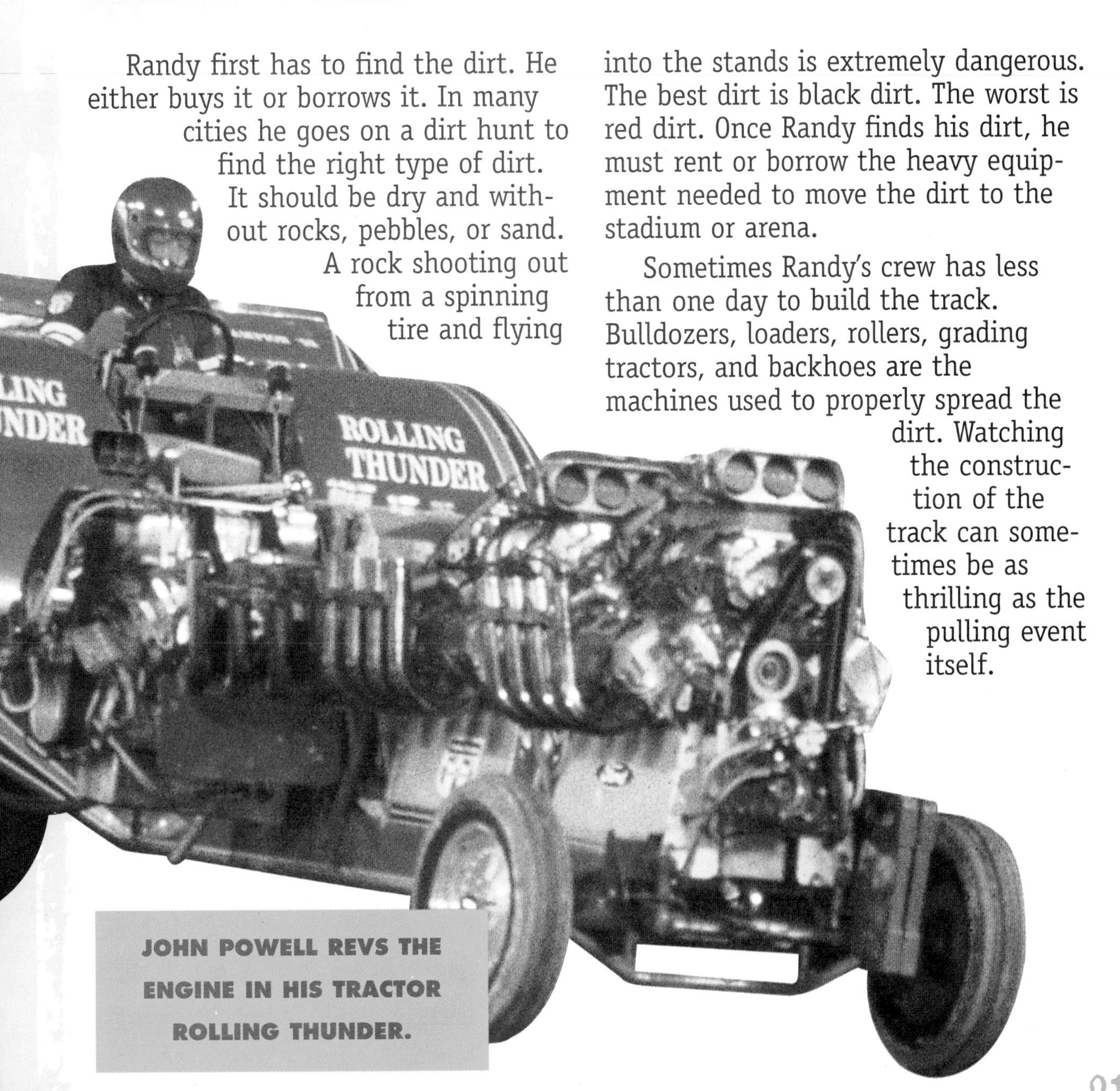

JOHN POWELL REVS THE ENGINE IN HIS TRACTOR ROLLING THUNDER.

CHAPTER 5

A Great Pioneer

Each year a few great athletes are presented the great honor of being inducted into their sport's Hall of Fame. This is usually done several years after the athlete retires. Art Arfons was inducted into the Motorsports Hall of Fame of America in June 1991. What is so amazing is that Art was still competing—and continues to do so today. Art started the 1995 tractor pulling season as a 20-year pulling veteran. It was another wonderful chapter in the incredible story of Art Arfons.

Art began his motorsports career back in 1952 when he became a drag racer. He quickly became the first racer to hit the 150 mile-per-hour mark and then the first to hit 200. He eventually got his drag racer to go a blazing 294 miles an hour in just 5.5 seconds. "The tracks were never long enough," Art remembers. "It was a real hassle to stop. You could crash your car or go off the end. I wrecked a few cars during my day, and that sure takes the fun out of it."

With dragsters going off the end of short tracks, something had to be done. Art came to the rescue by inventing the car parachute. Today, when a dragster crosses the finish line, a parachute opens from behind to stop the car.

Art often raced on the salt flats in Utah, where he set several more records. He broke the world land-speed record several times—once going as fast as 610 miles per hour!

"That was a terrifying yet thrilling ride," he says.

Art quit drag racing in 1972. He figured his motorsports career was over. Two years later, his neighbor in Akron, Ohio, convinced him to go to a pulling event. Art was astounded by the powerful engines. "That's when I really fell in love with pulling," he says.

Art went right to work building his own tractor puller. He joined the competition the following year with the Green Monster. Switching sports can only be done by the most gifted athletes. Examples include basketball star Michael Jordan, who for a while switched to baseball, and Deion Sanders, who has played both baseball and football. Art is among this select few. In 1990, he captured the USHRA All American Pulling Series Championship for the third time in his career. Then he was inducted in the Hall of Fame. "That was very special, especially since they put me in while I am still alive," he says. "That way you know you made it."

LEGENDARY RACER ART ARFONS WAS ASTOUNDED AT THE POWERFUL ENGINES USED IN PULLING TRACTORS.

Art has contributed in many ways to the sport. He built the first **turbine tractor** and invented the **pull-fender** and the **wheelie bars**. And Art isn't the only member of the Arfons family competing on the tractor pulling circuit today. His daughter, Dusty, has been a puller since she was 18.

When Dusty graduated from high school in Akron, her father made her an offer. "Dusty," he said, "I'll pay for your college or build you a tractor. You make the choice."

Dusty did not hesitate with her answer. She took the tractor.

Chapter 6

Women Pullers

Dusty Arfons had an exciting childhood. She traveled across the country in the family bus every year from the time she was eight years old. She would miss about 25 school days each year, touring parts of the country other kids her age were only reading about in their geography books.

Dusty always wanted to be a tractor puller like her father—the legendary Art Arfons. When she graduated from high school and Art offered to build her a tractor puller, Dusty screamed with delight. Her brother is Turbo Tim Arfons, a jet-powered-motorcycle jumper. Dusty was thrilled to be joining the family in motorsports competition. But would she be good enough?

There must be something in the Arfons family blood. Dusty joined the pulling circuit in 1985 and was an instant star. As the only woman competing, she finished fourth overall in the yearly point standings. She registered another top ten finish her second year and then another and another. Men who at first were bitter about losing to a woman began showing Dusty the respect she deserved. Going into the 1995 season, Dusty was one of the most feared competitors around. She has finished in the top ten every single year since she started. "Dusty has the competitive fires of the Arfons family, so she isn't easy to beat," says veteran

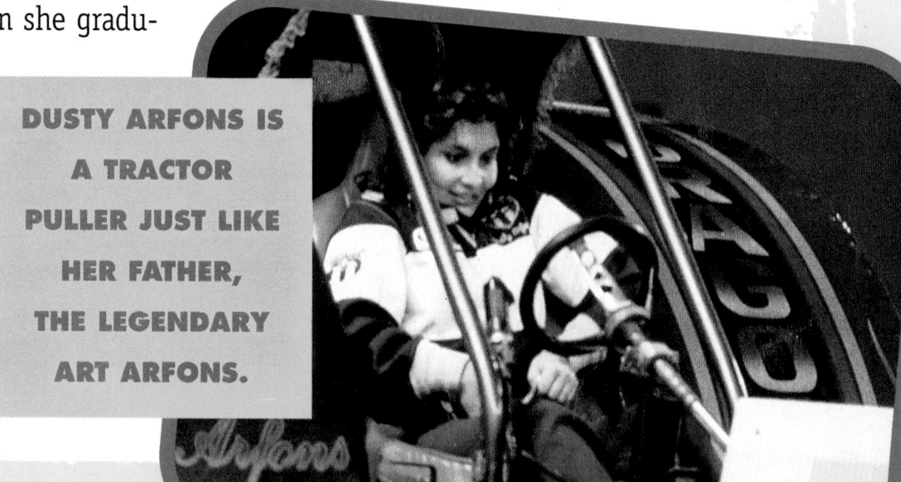

DUSTY ARFONS IS A TRACTOR PULLER JUST LIKE HER FATHER, THE LEGENDARY ART ARFONS.

THE DRAGON LADY SPITS FIRE AS DUSTY POWERS DOWN THE TRACK.

driver John Hileman. "We don't cut her any slack."

Dusty's nickname is Dragon Lady, and that also is the name of her tractor puller. "I have a lot of fun competing, especially when it's against Dad," Dusty says. "It's fun going against him because if I beat him, I know I really did something special."

Dusty remains the only woman tractor puller, but she is not the only woman competing on the circuit. Donna Webb is a truck puller. Strangely enough, Donna began competing in 1985—the same year Dusty started.

Donna fell in love with pulling the day she and her husband, Freddy, joined a pulling club in their hometown of Claremore, Oklahoma. She was overcome by the raw power of the massive engines in pulling trucks.

Donna is a big fan of actor Clint Eastwood, so she named her truck Sudden Impact after one of Eastwood's movies. Donna works in Claremore as an office manager. Freddy hauls Sudden Impact from venue to venue around the country, and Donna flies to the event on the weekends to drive the truck. She routinely finishes high in the yearly point standings and has won several events, including a victory at the 1989 finals in Houston.

"Pulling takes finesse and concentration, and Donna does it just like anyone else," STP driver DeWayne Baethge says. "Donna becomes very tough when it's all on the line."

STP DRIVER DEWAYNE BAETHGE SAYS PULLING TAKES FINESSE AND CONCENTRATION.

Chapter 7

Mastering the Fine Points

It isn't just driving skill that makes a great truck or tractor puller. Perfecting the fine points of the sport is what separates the champions from the rest. Many things are involved in pulling that the average fan doesn't see—things like selecting the proper **gear ratio**, calculating the correct **tire pressure**, and reading the track by examining the mixture of sand, clay, and water. Professional pulling truly is a complicated business.

Glenn Davis is one of the few competitors who has mastered all of the fine points. Glenn is known as Mr. Sparkle, and his expertise shines bright.

Glenn is recognized nationwide in his four-wheel-drive 1988 Chevy S-10 with its 1,500-horsepower engine. He has driven his shiny blue supertruck to five consecutive four-wheel-drive national pulling titles. No one else in pulling history has recorded such a winning streak.

Roaring through the USHRA field each week is no surprise to Glenn. He won competitions from the moment he took up the sport. Glenn started racing in 1981 in his hometown of Hollywood, Florida. He discovered secrets of the business by talking with fellow competitors, learning early that each event is unique. He then prepared for each race by paying close attention to detail. Glenn won the Florida state championships the next two years, and everyone was wondering, "Who is this Mr. Sparkle?"

Starting in 1984, Glenn took his truck to events throughout the South. In the next five years, he won eight regional championships and two national titles. A lot of the drivers were skilled. But Glenn was different. He didn't just jump in his truck and push the **accelerator** to the floor. First he figured out a strategy. He thought about things.

After winning his fifth straight national title at the age of 31, Glenn decided not to return to the competition in 1994. "Maybe I'm afraid of ending up second or third," he said with a laugh. "The streak is going to stop sooner or later. When you're number one, there is only one way to go and that's down."

The winners of the 1993 USHRA Tractor Pulling Championships in Bowling Green, Ohio, received rings for the first time. But Mr. Sparkle didn't win there. "To me," Glenn says, "winning a ring in pulling is like winning a Super Bowl ring in football. You have something to show people when you go out."

THERE IS MORE TO THE SPORT OF PULLING THAN JUST PUSHING THE ACCELERATOR TO THE FLOOR.

CHAPTER 8

Bowling Green

The campground is full. Tents are everywhere. Barbecue smoke floats through the air. Radios blare music. The pulling fans are here again. They play catch or pitch horseshoes or sit in their lounge chairs sipping beverages from coolers that are shaped like trucks and tractors. It's one big party—just as it is every year.

Bowling Green is the craziest, zaniest, wildest, most electrifying truck- and tractor-pulling event in the world. Thousands of people come early in the week to attend this week-end event in northwestern Ohio—and they always see the best the sport has to offer.

FANS ENJOY SEEING ENGINES LIKE THIS ONE CLOSE UP.

Larry Koester lost his lower legs in a farm accident. He knows all about being challenged. Larry comes every year to Bowling Green for what he says is his biggest challenge. He competes on the pulling circuit with the minirod tractor Footloose. Larry's stiffest competition is at Bowling Green. "If I could go to only one pulling event, it would be Bowling Green," he says. "I don't know what life would be like without all of the friendships I have developed here."

Larry's emotion for Bowling Green reached its climax in 1992 when he captured a title Friday night and took a third-place finish the next night.

The Northwestern Ohio Tractor Pullers Association is also known as the blue shirts. In 1967 the blue shirts began to create the Bowling Green Competition. They never imagined that it would eventually become the biggest outdoor pulling event in the world. Only 5,200

TINKER TOY COUGHS UP THE DIRT AS IT HAULS THE SLED DOWN THE TRACK.

fans and a handful of drivers showed up that first year. Five years later, the audience had grown to 25,000. Today more than 70,000 fans pour through the gates to watch more than 250 pullers. The pit is open, so the fans can get a close look at the vehicles and talk with their favorite drivers. Two-wheel-drive champion Ken Lamont says, "You can't fool these people. They are die-hard fans, and they know more about me than I do."

The rowdier fans gather in what is known as the animal section, while the more peaceful fans and families gather at the opposite end of the track. The animal section is filled with supersoaker water guns, beach balls, water hoses, and rollicking good fun. But the other side is just as boisterous when it comes to cheering for the pullers.

Ann Hileman wrote an article that appeared in the 1992 program for the National Tractor Pulling Championships at Bowling Green. Her husband, veteran driver John Hileman, had been trying—unsuccessfully—to win at Bowling Green for 19 years. John's truck, Golden Thunder, had won at practically every other venue throughout the country. Ann wrote, "There is just no frustration in pulling greater than winning every other event, only to have the big one elude you. It doesn't matter that you have a house full of trophies and a ship full of tool boxes because if you don't make it into the winner's circle at Bowling Green, you have suffered a big disappointment."

Maybe it was Ann's article that did it. Maybe it was just John's turn. Whatever it was, John Hileman won the two-wheel-drive championship that year at Bowling Green! And just for good measure, he won it again the following year! "I got a standing ovation from the crowd when I won it again," John proudly remembers. "Bowling Green is my favorite place on the circuit."

Chapter 9

Teamwork

You could be the best truck or tractor driver alive and still not win a single pulling event. That's because it takes more than driving skill to be a champion. It takes teamwork.

There is only one thing more important than the driver—the engine. That is what makes the **mechanic** so vital to a team. A good driver and his mechanic are capable of tearing down and rebuilding an engine in four hours or less. Most teams carry spare engines with them in case one breaks down during a pull. Still, there is plenty of tinkering with tools between pulls. It's a lot like performing emergency surgery.

Perhaps the best team today is Midnight Express, with driver Ken Lamont and crew chief Tom Murk. Twice they have been voted by fellow competitors as the Two-wheel-drive Team of the Year. "The two of them make a great team," driver John Hileman says. "Tom takes care of the engine and **chassis** to make them first-rate, and this gives Ken the time he needs to study the track, concentrate on driving, and do what he does best."

Ken has been competing since 1972. In high school he was a star trumpet player, ready to go to college with a full music scholarship. But at the last minute, Ken decided he wanted something different. He began driving truck pullers. "When I get in my truck, it's a real feeling of power!" he says. "The most important thing is to have a good machine. If we have a problem with it, Tom and I work on it until it's fixed. We stay up all night if we have to."

KEN LAMONT AND TOM MURK HAVE WON THE TWO-WHEEL-DRIVE TEAM OF THE YEAR AWARD TWICE WITH MIDNIGHT EXPRESS.

THE POWERFUL YELLOW TRUCK SUNDANCE KID HAS WON THREE STRAIGHT NATIONAL CHAMPIONSHIPS.

Tom's knowledge of engines and Ken's driving skills have combined to give red and silver Midnight Express over 100 pulling victories, several national titles, and a record 19 full pulls in one season. "Ken and I eat, sleep, and breathe this stuff," Tom says. "To be able to spot the problem, fix it, and do better the next time is what creates a winner."

There are dozens of winning teams on the circuit today. They each have a story to tell, a reason for racing, a special moment in their career, and a powerful truck or tractor.

Ray Carpenter powered his way to back-to-back national pulling championships in the two-wheel-drive competition. He was tearing up the competition again in 1993 when the

Mississippi River flooded his hometown of Bettendorf, Iowa. Ray put his competition on hold to tend to the family business during the flood. A young man named Danny Hubert offered to drive Ray's yellow truck, Sundance Kid, while Ray was busy. Ray handed Danny the keys and told him to give it his best shot. Danny did more than that. He went on to capture the 1993 national points series. Now Danny is Ray's full-time teammate.

"The first time, they call you lucky. The second time, they say it's a fluke. The third time, they know we're good," Ray says about winning three straight championships with Sundance Kid.

There is plenty of two-wheel-drive competition for Ray Carpenter and Danny Hubert. Skyline Skreamer, driven by Curt Poole, consistently finishes in the top ten, as does Spike, with driver Bill Humphrey; Rare Breed, with driver Darrell Varner; and Just For Fun, driven by Rich Santefort.

A PULLING ENGINE HAS TO BE STRONG ENOUGH TO WITHSTAND THOUSANDS OF POUNDS OF RESISTANCE.

SATURDAY NITE DRILLER ALWAYS POSES A THREAT IN THE FOUR-WHEEL-DRIVE COMPETITION.

Among the top four-wheel-drive pullers are Saturday Nite Driller, with driver Tommy Holman; High Roller, with Howard Lewis; Golden Ox, driven by Bud Jaske; Fastbreak, with driver Gary Varner; Tinker Toy, with driver Bruce McClintock; Mad Dog, driven by Aaron Mobley; Hillbilly, with Red Grogan; Milky Way, driven by Kenneth Popham; Outlaw, with driver Russell DeForrest; Rampage, with Harry Owens; Orange Krush, with Ron Brummel; and the ever-popular 1940 black Ford known as

WAR WAGON IS A MIGHTY TRACTOR WITH DRIVER PAUL NORMAN ABOARD.

Back in Time, with driver David Willoughby.

Some of the top tractors are War Wagon, with driver Paul Norman; Dirtslinger, with Bill Leischner; Rambunctious, with driver Duane Bergman; New & Improved Rambunctious, with Milt Bergman; Fancy Farmer, driven by Robert Elliott; Just Add Dirt, with driver Mike Piper; and Chevy Thunder, driven by Joe Eder, who used to play with tractors in his sandbox and dream about pulling.

Chapter 10

The Force to Pull

Of the hundreds of professional pullers today, none got their start as early as Jeff McPherson. Jeff's father, Richard, was competing before Jeff was born. Jeff went to his first pull when he was just eight months old.

Now Jeff and his father both compete. Richard drives Missouri Raider, the modified 1923 blue Ford Model T that fans everywhere love to see. Jeff drives Bean Bandit, a 1990 green and white Ford. Jeff hopes to have children to pass the tradition on to.

Another pulling family is the Walsh brothers from Mauston, Wisconsin. Dan Walsh owns the two-wheel-drive truck Irish Challenger. Dave competes with his two-wheeler Irish T. Both have won numerous contests, and Dan won the USHRA points championship in 1991. But the best vehicle of all is Dave's tractor, also known as Irish Challenger.

Dave's slick green tractor led the 1993 points standings when he arrived with it at Bowling Green for the final event of the year. But on Friday night, disaster struck. The teeth on the crank gear sheared off, causing $12,000 worth of damage

RICHARD MCPHERSON POWERS FAN FAVORITE MISSOURI RAIDER TO ANOTHER FULL PULL.

to the gear heads. With two days of pulling left, the points title and the massive winner's check that goes with it were in jeopardy. Dave and brother Dan stayed up all night reassembling the engine, finishing just in time for the Saturday afternoon event. Irish Challenger captured second place that day and then finished third on Sunday, propelling Dave to the national points title.

The force that drives people to become truck and tractor drivers is strong—very strong. How else does one explain Dr. Wayne Rausch? Yes, that's *Doctor* Wayne Rausch. Wayne was a respected professor at Ohio State University earning a healthy income. He gave up that career to become a puller. Wayne powers through the mud with Yellow Model T and Little Red Truck—a pair of two-wheel-drive trucks that have captured nearly every title there is. Wayne has won dozens of national events. He has won the points series. He has been chosen as Puller of the Year. But his favorite award of all was being honored with his crew chief as Team of the Year. And who is Wayne's crew chief? His wife, Jo.

Yes, the force to become a puller is very strong indeed.

EVEN DOCTORS ENJOY PULLING. PROFESSOR WAYNE RAUSH POWERS LITTLE RED TRUCK TO VICTORY.

Irish
Challenger

THE IRISH CHALLENGER NEEDED ROUND-THE-CLOCK REPAIR WORK AT BOWLING GREEN TO WIN THE NATIONAL TITLE.

Glossary

accelerator The foot pedal used to control the speed of a vehicle

chassis The frame, wheels, and machinery that support a truck or tractor

four-wheel drive A type of vehicle in which all four wheels are powered by the engine

full-pull The pulling of a sled the entire length of the track

gear ratio The rotational speeds of the gears that determine speed and power

hitch The metal device that connects the sled to the pulling vehicle

horsepower The way the power of an engine is measured

kill switch A device that, when pulled by the sled operator, automatically shuts off the engine

mechanic The person who performs repairs and fine tuning on the truck

minirod A small-sized tractor

modified A term used to describe an engine that has been changed to make it more powerful

pull-fender A fender designed to connect to a pulling sled

pull-off A contest in which two or more pullers compete a second time to break a tie

skid plate The flat piece of steel at the front of the sled that digs into the ground

step-on sled A sled used in the 1960s upon which men would step to add their weight to be pulled by a tractor

tire pressure The amount of air in a tire, which determines its firmness

tug pull A competition that took place in the 1930s in which tractors pulled heavy objects

turbine tractor A tractor that runs on a water- or steam-driven engine

two-wheel drive A type of vehicle in which only the rear wheels are powered by the engine

weight transfer machine (also called a sled) The heavy object that is pulled by trucks and tractors in competitions today

wheelie bars Metal poles that prevent a truck operator from popping a wheelie

DIRT

INDEX